The contemporary relevance of the social sciences

ACADEMY of SOCIAL SCIENCES

CAMPAIGN for SOCIAL SCIENCE

The Campaign for Social Science demonstrates how social science improves public policy, society and all our lives. It highlights the value of applied social science research and advocates for its greater use in decision-making and in government. We promote the benefits of investing in social science to deliver evidence-based solutions to the significant challenges and opportunities facing our society. The Campaign is supported by a coalition of universities, social science societies and publishers.

The Campaign is part of the Academy of Social Sciences – the only body in the UK that exists solely to promote the social sciences sector and to articulate the social sciences' importance for public benefit. We are a national academy and the UK professional body for academics, practitioners and learned societies in the social sciences. Our work is informed and supported by 1,700 leading social scientist Fellows together with 48 member Learned Societies that cover the main disciplines and subdisciplines in the social science sector. This gives us a reach of around 90,000 social scientists in the UK.

The social science disciplines include: anthropology, architecture and planning, business and management, criminology, development studies, economics, education, human geography and environmental studies, law, politics and international studies, psychology and behavioural sciences, regional studies, sociology, social work and social policy.

To contact us, please email:
media@acss.org.uk for media enquiries
office@acss.org.uk for general enquiries
Or call +44 (0) 300 303 3513

For further information, see:
www.acss.org.uk
X: @AcadSocSciences and @CfSocialScience
LinkedIn: www.linkedin.com/company/academy-of-social-sciences/

The Academy of Social Sciences is a company registered in England, number 3847936, and a registered charity, number 1088537

ACADEMY of SOCIAL SCIENCES

CAMPAIGN for SOCIAL SCIENCE

The contemporary relevance of the social sciences

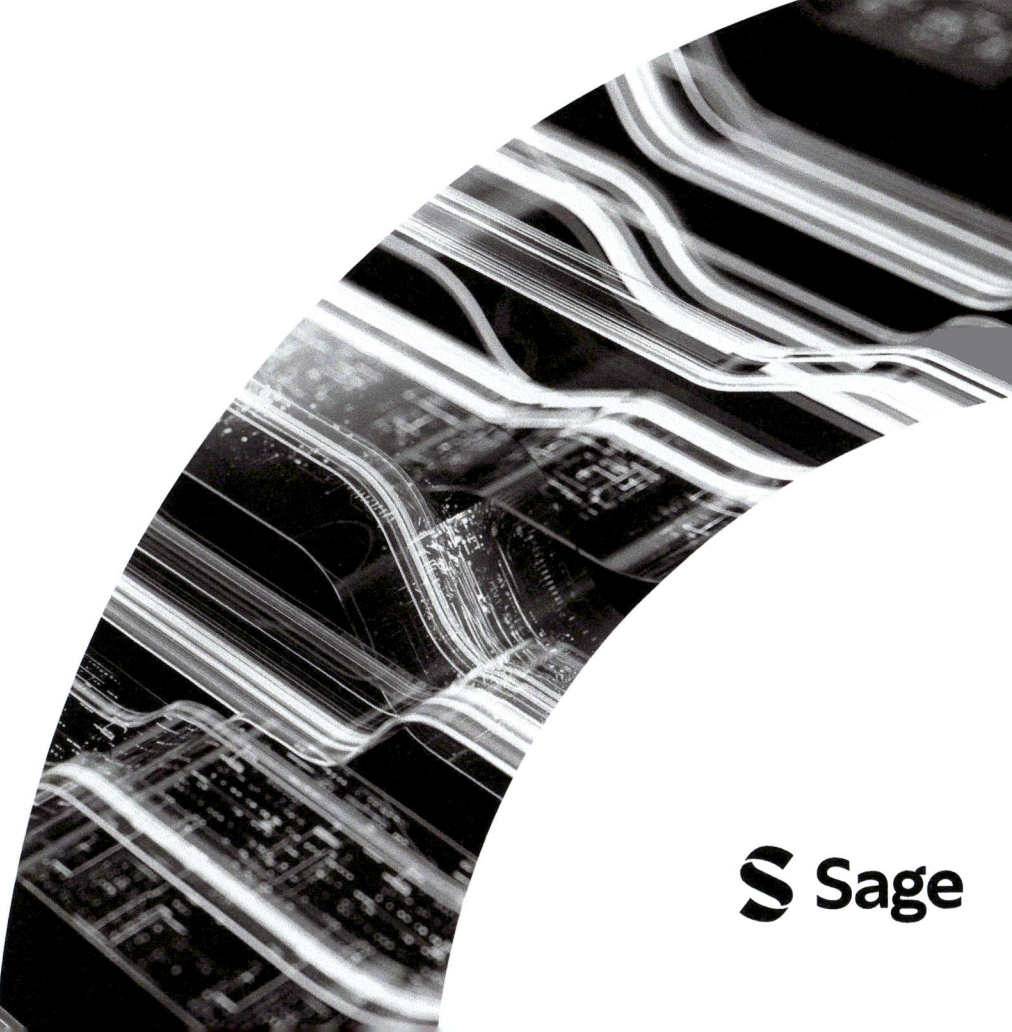

S Sage

ACADEMY of SOCIAL SCIENCES | CAMPAIGN for SOCIAL SCIENCE

Academy of Social Sciences
c/o Knox Cropper LLP
5 Floor, 65 Leadenhall Street
London, EC3A 2AD
+44 (0) 300 303 3513

1 Oliver's Yard
55 City Road
London EC1Y 1SP

2455 Teller Road
Thousand Oaks
California 91320

Unit No 323-333,
Third Floor, F-Block
International Trade Tower,
Nehru Place
New Delhi 110 019

8 Marina View Suite 43-053
Asia Square Tower 1
Singapore 018960

British Library Cataloguing in Publication data

A catalogue record for this book is available from the British Library

ISBN: 978-1-03-624144-5
DOI: 10.4135/wp521103

BB0357021

Printed and bound in Great Britain by Bell and Bain Ltd, Glasgow

Contents

Foreword

This report marks the culmination of a two-year project filling a vitally important gap. It lays out the rock-solid evidence of the essential role of social science in harnessing the key opportunities and mitigating the many and varied challenges we face in our contemporary human world.

We have a hugely successful and world-leading social science sector in the UK, but as a community we have been too reticent and fragmented in making that case. We have a reputation for excelling at pointing to problems, which is, of course, a vital role – but so too is our ability to find solutions. This reputation of only putting forward problems is unfair but also, perhaps, understandable; we have focussed insufficiently on making a strong and sustained case for our essential positive contribution of solution-seeking.

This report attempts to address this, through a mix of key facts and compelling stories that we all should take pride in and, more importantly, communicate relentlessly. This is not for its own sake or to make us feel better about our work – we truly believe that social science evidence and expertise improves decision-making, society and lives, so making this case serves a clear, wider purpose.

Here, we draw together and build on some of our other recent reports[1] and translate that into a simple narrative about the value and contemporary relevance of our social sciences. This report has the weight and depth of previous work underpinning it, but it makes the bigger case that, for all the reasons set out in those individual reports, now is the time to recognise the

---

[1] These include, but are not limited to:
- Beyond the ballot: social science insights on eight key policy challenges (Nov 2024)
- Research funding in the UK social sciences: summary data report for 2013/14 to 2021/22 (Jul 2024)
- Equality, diversity and inclusion (EDI) in the social sciences: summary data report (Mar 2024) – with the ESRC
- The SHAPE of research impact (Jan 2024) – with the British Academy
- Reimagining the recipe for research and innovation: the secret sauce of social science (Jan 2024)
- Social sciences in a time of change, 2020–2022 (Jul 2022)
- The place to be: how social sciences are helping to improve places in the UK (Nov 2021)
- Vital business: the essential role of the social sciences in the UK private sector (Sep 2020)
- Positive prospects: careers for social science graduates and why number and data skills matter (May 2019)

vital role of the social sciences in rising to the challenges of the modern world.

The report can be read as a stand-alone, but our objective is to effect change, so it will be accompanied by a range of additional resources to bring it to life, including case study examples for you to use, and re-use. The report is a culmination of much work, but it is only the start of a campaign which we will ensure evolves and adapts. We'd very much like to hear your ideas on adding to the case, and your help in ensuring its impact.

It is also part of a wider ongoing programme, where each strand reinforces the other. Key amongst these is another current project examining the evidence infrastructure in government, making constructive and pragmatic recommendations for how it could be improved. The cases made in this report for how vitally important the social sciences are as part of that evidence base helps explain why government should engage more fully, beyond the current pockets of excellence.

The report is built on the excellent work of the core Campaign for Social Science team, our Board and, most of all, our supporters, in Sage and our universities. None of this would be possible without their incredible insight and commitment to ensuring the social sciences are recognised as not just important but essential.

Professor Bobby Duffy FAcSS
Chair, Campaign for Social Science

Why the social sciences matter: eight principles

Which group of disciplines informs us about the dynamics of our changing human world and helps us understand and manage many of the issues facing society, the environment, the economy and places, in the UK and internationally? Which group of disciplines also makes a positive contribution to our daily lives – through the economic value they provide, the tangible differences they make to society, the innovation they facilitate and through guiding good policy decisions? And which group of disciplines offers opportunities to all, through the skills and knowledge they afford and the diversity of careers to which they can lead?

The social sciences uniquely meet all three key criteria.

This report explores these points in outlining eight principles about the UK's social science sector, underscoring the discipline's contemporary relevance and importance to modern British society and the wider world:

1. The social sciences are **the understanding of people, society, economy and places**

2. The UK is a **social science powerhouse**

3. The social sciences bring **economic value**

4. The social sciences **serve society**

5. The social sciences **facilitate innovation**

6. The social sciences **improve people's lives** by guiding good decision-making

7. Social science subjects are open and **available to all**

8. Social science skills are **valued by employers** and lead to well-paid careers

The report sets out each principle in detail, along with some supporting evidence. It is written as concisely as possible so it can remain a user-friendly reference, with footnotes and links to underpinning research, reports, datasets and studies that substantiate and illuminate the eight principles. There is, of course, a substantial breadth and depth of alternative case studies sitting behind each principle – necessarily, we have picked two strong examples for each section. Likewise, at times we have made generalities to keep the text concise and readable.

In setting out the principles, the report is designed to complement our dedicated web pages on why social science matters. The web pages are written for a public audience, whereas this report is aimed primarily at policymakers and civil society.

We hope this report is instructive in setting out the strength, reach and relevance of the UK's world-leading social science sector, but also that it can be a prompt for dialogue about how to make the most of the disciplines' potential in the years ahead. The Campaign for Social Science will continue to foster that dialogue among the academic, practitioner and user communities.

There is nothing more important to the survival of society than social science and the humanities.

STEM Panel Member, REF 2021

Principle 1: The social sciences are the understanding of people, society, economy and places

Social science is the study of people – as individuals, communities and societies – and their behaviours and interactions with each other and with their built, technological and natural environments. Social science seeks to understand the evolving human systems across our increasingly complex world and how our planet can be more sustainably managed. It's vital to our shared future.

Social science includes many different areas of study, such as how people organise and govern themselves, and broker power and international relations; how wealth is generated, economies develop, and economic futures are modelled; how business works and what a sustainable future means; the ways in which populations are changing, and issues of unemployment, deprivation and inequality; and how these social, cultural and economic dynamics vary in different places, with different outcomes. This understanding sits at the heart of people's lives, the UK's future and policymaking at a UK Government and devolved government level.

The social sciences have a critical contribution to make in helping us understand, imagine, and craft a more sustainable future for all.

UNESCO

As this report sets out, the social sciences are:
- **Popular**: Of the 12 most popular A Level subjects, five are social sciences: business studies, economics, geography, psychology and sociology.
- **In demand**: Around 40% (>230,000) of new undergraduate students chose to study social science subjects at UK universities in 2020/21.
- **Well paid**: Social science graduates are well paid. Many earn as much as STEM graduates.
- **Vital to business**: Social science knowledge and skills, including leadership and strategic planning, international liaison, sustainability, consumer growth, marketing, legal and financial management, are vital to business operations.

Image on p7 used under license from Shutterstock.com

- **Relevant to governments**: Around 80% of the UK Government's stated Areas of Research Interest (ARIs) are questions which require social science expertise and evidence.
- **Inspiring**: Social science helps us make sense of our changing human world, inspiring us to be culturally sensitive, environmentally aware and socially responsible citizens.

The value of applied social science research

Social science research can define and diagnose the major challenges facing society, identify the policy levers to address them, and help inform interventions to implement change.[2] Furthermore, social scientists are adept at working across disciplinary boundaries to tackle societal problems[3] – an inherent requirement of the new UK Government's 'mission-led' approach to policymaking. Yet the social sciences' impact is felt beyond government policy; it is at the forefront of providing tangible benefits to the lives of citizens. In addition to the two case studies (p11 and p12), other examples include:

- **Recovering teaching disrupted during COVID-19**: COVID-19 lockdowns saw 24,000 schools in England alone forced to close for significant amounts of time, affecting the learning of hundreds of thousands of children. A team from the University of Oxford's Department of Education has been investigating the effects of the Education Endowment Foundation's (EEF) interventions on protection from COVID-19 learning loss. Amongst other things, the study found that the gap in maths attainment between disadvantaged students and others has widened since the COVID-19 pandemic began – increasing by approximately 4% to 17% compared to before the pandemic, equivalent to about one month's progress.

---

[2] Our own recent report, Beyond the ballot: social science insights on eight key policy challenges, emphasises this by summarising contributions from over 100 leading social scientists on contemporary policy issues.

[3] Wilsdon, J., Weber-Boer, K., Wastl, J. & Bridges, E. (2023) Reimagining the recipe for research & innovation: the secret sauce of social science, London: Sage/Academy of Social Sciences.

Image used under license from Shutterstock.com

- **Transforming everyday mobility patterns in a divided city**: Belfast has a population of around 350,000 people, 48% of whom identify as Catholic and 36% as Protestant. Mapping out the everyday movements of citizens in this divided city has given policymakers a better understanding of the psychology of sectarian divisions. This has shaped the approaches of Belfast City Council and led to the encouragement of more residential accommodation in the city centre, along with the creation of a new campus site for Ulster University to support this.

- **Recovering unpaid wages**: Research detecting and improving the recovery of unpaid wages showed that wages amounting to £3.1bn annually were being left unpaid in the UK. These findings led to better regulation and enforcement of entitlements to holiday pay as well as to greater efforts to pursue directors evading employment obligations. Improved enforcement directly benefited over two million of Britain's lowest paid workers.

- **Improving quality of life for dementia patients**: Research into the therapeutic value of playful objects led to the development of a product that improves the quality of life of people affected by advanced dementia (over 100,000 people across the UK), which is now prescribed on the NHS. A six-month trial funded by £185,000 from Welsh Government found the product improved wellbeing for 87% of participants. National and international demand followed from Health Boards, care homes and the public. As a result, a spin-out company was launched in 2020, backed by investment capital from sources including a crowdfunding campaign and Alzheimer's Society.

Case study: Born in Bradford

Social science is <u>making a difference to the health of communities</u>. Born in Bradford is tracking the lives of over 40,000 people to help improve the health and wellbeing of the local community. Its evidence has led to projects including improving urban green spaces, an early life intervention programme and establishing a clean air zone in Bradford.

This is all about how we can improve the services within our city and how we can make things better for families.

Professor Rosie McEachan
Director, Born in Bradford

Image used under license from Shutterstock.com

Case study: City-REDI

Social science is making a difference to the economic resilience of the UK's nations and regions. Research by City-REDI produced accurate projections of the effects of Brexit and COVID-19 on the UK economy. This directly informed the West Midlands region's socio-economic policy and led to the region receiving £1.5bn to accelerate planned infrastructure projects to inject extra money into the economy and create new jobs.

This [a 50% uplift to funding in the region] is a result of the robust, evidence-based case which the WMCA put forward, with City-REDI analysis at its heart.

Head of Economy and Local Industrial Strategy, West Midlands Combined Authority

Image used under license from Shutterstock.com

Principle 2: The UK is a social science powerhouse

Overview

UK social science delivers tangible real-life impact at local, national and global scales. Moreover, it is an area of genuine competitive strength for the nation, with our social science departments and research being the envy of many parts of the world. UK universities comprise one-fifth of the top 50 social science universities in the world (including three in the top ten),[4] but the success of our social sciences has come against a challenging financial backdrop – for universities as a whole but specifically for social science research compared to some other parts of the academic sector.

Social science is poorly-structured and underfunded in a UK context, which makes it harder to get research done at scale, develop and provide evidence bases that are produced with rigour and can be used with confidence, and to attract and keep the researchers who can make breakthroughs in relevant fields to improve society. Addressing these areas is essential if the UK is to retain its status as a world leader in cutting-edge social science research, of the types illustrated in the case studies overleaf which bring tangible societal benefits.

The value of UK social sciences

The UK as a world leader in high impact social science research is evidenced in the 2021 Research Excellence Framework (REF) exercise: 80% of social science research was rated either world leading (37%) or internationally excellent (43%).

> **Did you know ...?**
>
> UK universities are among the top ranked in the world for social science, with ten UK-based universities being ranked in the top 50 universities in the world for social science. Three of these are ranked in the top ten.

A recent report[5] commissioned by the Academy of Social Sciences and the British Academy further exemplifies the social sciences' value for money and impact. Social science is fundamental to understanding and helping mitigate many of the economic, social, place-based and environmental challenges we face in the UK, and in contributing to multidisciplinary 'missions'.

---

[4] QS University Rankings (2024) QS world university rankings by subject 2024: social sciences & management, QS/Top Universities website.
[5] Wilsdon, J., Weber-Boer, K., Wastl, J. and Bridges, E. (2023) Reimagining the recipe for research and innovation: the secret sauce of social science, London: Sage/Academy of Social Sciences.
Image on p13 used under license from Shutterstock.com

We also know the social sciences are delivering insights for critical areas of public policy. Perhaps most notably, social science research informed many dimensions of policy and practice in managing the COVID-19 pandemic, whilst a database of research priorities recently published by the UK Government Office of Science is dominated by 'social science' questions.

Because of these strengths, and as examined in Principle 8, the social sciences are subjects which students want to study, which serve society well, and which are valued by employers.

Social sciences' funding relative to other sectors

Our report[6] published in summer 2024 set out Higher Education Statistics Authority (HESA) research funding data for the nine academic years between 2013/14 and 2021/22, to document the scale and trends in research funding for the social sciences across UK higher education. It analyses available data to compare the social sciences with the three other academic sectors: the medical and biological sciences; arts and humanities; and science, technology, engineering and maths (colloquially referred to as STEM).

As well as logging change over time in research funding, it also raises important questions about what the UK wants from its social science research and the appetite to fund it. The funding differential in real terms between social science research and the equivalents from STEM and the medical and biological sciences has grown ever wider, despite a record performance from social science research and impact, and unprecedented need for the insights it provides. This is a supply constraint, not a demand constraint.

[6] Academy of Social Sciences (2024) Research funding in the UK social sciences: summary data report for 2013/14 to 2021/22, London: Academy of Social Sciences.

Image used under license from Shutterstock.com

Recommendations

Within this context, we argue that the UK Government needs to consider whether it is getting as much benefit as it might out of our world-leading social science research base. To fully harness its potential, and based on the data analysis and contextual changes over the past nine years, the Campaign for Social Science recommends the UK Government and UK Research and Innovation (UKRI):

- Review urgently the adequacy of the research funding levels for the social sciences sector, including their involvement in multidisciplinary, challenge-led research.
- Consider specific additional funding of education research, research that benefits a specific nation/region of the UK, and research in the social aspects of health sciences because of their centrality to delivering on the UK Government's stated policy objectives.
- Sustain the UK's involvement in the EU Horizon programme because of the essential research collaborations and economic growth this supports.
- Ensure a sound balance of 'blue skies' and applied research, alongside new conceptual thinking and novel methodological approaches, which is supported by UKRI and the ongoing contribution of QR funding, to enable such research to underpin innovation and future applications.

In doing so, we believe the UK would release more of the social sciences' potential to contribute to the social, economic and environmental wellbeing of the nation, and start to understand and address many of the challenges we face across the nations and regions of the UK.

Case study: Monitoring crop yields from space

Social science, as part of broader multidisciplinary research across different sectors, is making a difference to food security through improving vegetation monitoring. Social scientists developed a series of algorithms which estimate the chlorophyll content of land-based vegetation in near real time. This significantly enhanced the ability of the European Space Agency and European Commission to monitor and map global vegetation from space, generating huge benefits for providing timely and targeted responses to poor harvests and plant disease outbreaks.

You can use the dataset to estimate the amount of carbon being captured by vegetation and hence we can look into longer impact of climate on our ecosystem.

Professor Jadu Dash

Image used under license from Shutterstock.com

Case study: Understanding Society

Social science is <u>making a difference to improving our understanding of life in the UK</u>. The 'Understanding Society' study is one of the largest long-term panel studies in the world, drawing on quantitative data from around 40,000 UK households. The 'big data' arising from it is used by researchers, charities, businesses and UK Government departments to understand and inform solutions to societal issues relating to education, employment, health and wellbeing, politics and social attitudes, immigration, transport and environment, and young people.

Understanding Society is an important resource for tracking trends in children's wellbeing in the UK over time … it helps us identify areas of children's lives that need attention.

The Children's Society

Image used under license from Shutterstock.com

Principle 3: The social sciences bring economic value

Overview

Social science is fundamental to business and economic development, providing value through:

- **Targeted research knowledge** – both undertaking and translating research to inform business development and business innovation.
- **Professional skills** – underpinning the core operations of all businesses, including finance, legal affairs and marketing.
- **Socio-economic and geopolitical contextualisation** – understanding economic and place-based settings, changing demographics, attitudes to work, and trade and legal policies, which collectively are necessary for efficient business management and economic growth.

The social sciences' contribution to the economy

The contribution of the social sciences through research was clearly demonstrated in work carried out last year[7] analysing impact case studies from the 2021 REF exercise. This showed that social science research is bolstering UK expertise and strength in areas of competitive advantage, including driving innovation in banking and finance, and in legal services. One whole subcategory of impact uncovered through the analysis is dedicated to macroeconomics and finance, including examples of research improving central bank operations and developing new approaches to financial forecasting.

Separately, the social sciences' contribution through the skills and services they fuel has a huge role to play in the UK economy. Companies rely on the skills and knowledge of social scientists to understand and engage with their markets, clients and consumers, to analyse and manage risk and long-term strategies, to develop new products or ways of working, and to engage in multidisciplinary ways of working. Examples can be found in the case studies (p23 and p24), and our own Vital Business report[8] also detailed some of the ways in which UK private-sector businesses value and incorporate social science knowledge and insights into how they run and grow their firms.

> **Did you know ...?**
> 80% of the UK's economic output in 2024 was accounted for by the services sector, which is fuelled by the social sciences.

---

[7] Wagner, S., Rahal, C., Spiers, A. et al (2024) The SHAPE of research impact, London: British Academy.

[8] Lenihan, A. & Witherspoon, S. (2020) Vital business: the essential role of the social sciences in the UK private sector, London: Academy of Social Sciences/Campaign for Social Science.

Image on p19 used under license from Shutterstock.com

Many companies use social science knowledge and skills in their research and development (R&D) for the longer term, investing in human capital and skills, but also in innovations that lead to new services or products.

Finally, the social sciences contribute to broader economic growth by providing an essential lens for industry, policymakers and broader civil society. Some of these issues are explored in greater depth in Principle 6, where we set out the importance of decision-makers being able to draw on robust insights and 'big data' around demographic, economic, societal, environmental and behavioural factors.

Quantifying the social sciences' value

The Academy of Social Sciences will shortly be commissioning work to establish the current value of the social sciences to the UK economy. This will be a major study, drawing on the best current data to establish social sciences' contributions across a range of indices.

By way of a guide, the most recent data, based on analysis[9] from 2014 and adjusted for inflation, would indicate the 2025 value to be in the region of £32.8bn. This comprises the collective economic value of social science teaching and research in UK universities (£4.8bn a year in 2014 figures) plus the costs that the financial sector, business corporations and public sector agencies spend on employing professional social scientists to mediate or translate academic research into their organisations (at least £19.4bn a year in 2014). These figures will obviously have changed significantly in the decade since the original data but are included here for indicative purposes. It should also be noted that these figures are an underestimate, as they do not include things like the economic value of professional social science skills and knowledge at the core of UK businesses – a critical strand of economic value given that the social sciences underpin the services sector, which in 2024 accounted for 80% of the UK's economic output.[10]

---

[9] Bastow, S., Dunleavy, P. & Tinkler, J. (2014) The impact of the social sciences: how academics and their research make a difference, London: Sage.
[10] Hutton, G. (2024) Economic update: which industries have grown in 2024?, House of Commons Library website.

Recommendations

To safeguard and build the social sciences' contribution to the UK economy, we recommend that:

- Research funding allocations be reviewed to reflect more fully the contribution the sector makes; and their substantial contribution be reflected in the forthcoming review of QR funding. It is also vital that the UK Government and its devolved counterparts ensure in the short-term that QR funding retains its real-term value.
- Social science-based R&D should be included within R&D tax relief at the earliest possible opportunity.
- The Frascati rules for measuring R&D should be expanded to reflect the social sciences' contribution. We would welcome an expanded definition focussing less on publications or patents, as recommended by the House of Commons Science & Technology Committee report[11] on balance and effectiveness in research and innovation spending.

---

[11] House of Commons Science & Technology Committee (2019) Balance and effectiveness of research and innovation spending, Westminster: UK Parliament.

Case study: Decision Maker Panel (DMP)

Social science is <u>making a difference to economic and business policy</u>. As part of the Bank of England's Decision Maker Panel, social scientists conduct a monthly survey of 9,500 senior business executives about current business conditions, expected future conditions and uncertainty. During COVID-19 this directly informed policy such as the furlough scheme and the easing of social restrictions. The Panel continues to provide monthly data and quarterly summaries used by the UK Government and the Bank of England.

Again and again the Committee has used the evidence provided by the DMP to guide its decisions. It is by far the most influential survey we use.

Monetary Policy Committee member

Image used under license from Shutterstock.com

Case study: SME growth challenges

Social science is <u>making a difference to small and medium enterprise growth</u> in the UK. In 2023, small and medium-sized enterprises (SMEs) in the UK accounted for over 99% of firms and employed 16.7 million people. Social science research has helped SMEs overcome the barriers they face to growth. Findings from social science have influenced policy to ease financial pressures on SMEs by the Scottish Government, including the Brexit Support Grant in 2019 and the Pivotal Enterprise Resilience Fund in 2020 which supported Scottish SMEs via grants totalling over £1.2 million.

[…Brown's] work fed into our thinking on discouraged borrowers, and more recently on the specific relationship with innovation will also help us better develop and target our policies.

Chief Statistician
BEIS, 2020

Image used under license from Shutterstock.com

Principle 4: The social sciences serve society

Overview

Social sciences provide the understanding of social and economic characteristics and contexts, the processes of change over time, and interconnections at all spatial scales. This is the bread and butter of social science – together with the data that informs it – which leads to ideas for solutions, interventions and sound implementation. In doing so, our disciplines deliver tangible improvements to people's daily lives – helping to reduce inequalities in education, supporting the tackling of health disparities, ensuring that people are paid fairly and so much more.

Furthermore, collaborative work between social scientists and researchers in other disciplines is often essential to finding multidisciplinary solutions to the most pressing global issues of our time. Insights from social science in areas typically seen as relating primarily to STEM, for example, have provided huge benefits in ensuring new technologies fit with the world around us. They have also provided a deeper understanding of people's thoughts and behaviour relating to the climate crisis.

The social sciences as society's 'glue'

Some of the most complex and pressing problems facing society are being tackled using research which draws heavily on social science insights.[13] Examples are seen within the healthcare sector, where impacts have improved mental health or informed methods and technologies for the detection, diagnosis and treatment of major health conditions. Elsewhere, behavioural, geographical and legal research are coming together to play a transformative role within sustainability and infrastructure. Significant areas of impact are also seen within the domains of education (with sociologists and psychologists playing a major role) and in government and law (with economics, political science and international development being prominent).

> ### Did you know ...?
> Social science research has directly informed changes to police practices, improving approaches in tackling rape and serious sexual assault offences.[12]

[12] Stanko, B. (2022) Operation Soteria Bluestone year one report, London: UK Government Home Office.

[13] Wagner, S., Rahal, C., Spiers, A. et al (2024) The SHAPE of research impact, London: British Academy.

This leads to real-world impacts of social sciences all around us. It can be seen in how our cities are designed, how we shape employment to not only deliver economic prosperity but also provide employment rights to workers, how we work with communities to mitigate and adapt to climate change, and how to deliver public services effectively and fairly so the sick can be healed, children can be given a proper education, society's most vulnerable can be provided with a safety net, and marginalised communities and individuals can be treated equitably. Examples across all these domains and more are on our website. Meanwhile, social science faculties and schools in our universities are major contributors to regional civic agendas, and in providing contextual understanding, data and insights into societal challenges at community and regional scales.

Social science research is happening all around us, with social science expertise spread across the UK in universities, business and industry.[14] This understanding of local circumstance and ability to tailor-make solutions to societal challenges at community level is an inherent strength of the social sciences. Any efforts to address disparities (social, economic, cultural, environmental) across the nations and regions of the UK necessarily involve a social science lens.

Finally, social scientists are adept at working across traditional disciplinary boundaries to understand and address complex issues for the benefit of society.[15] This was shown at pace when social scientists worked to inform our understanding of the effects of COVID-19.

For example, commencing around the start of the first lockdown, a team of researchers at UCL used online data from over 70,000 adults in the UK to track the psychological and social consequences of the pandemic. This example of social scientists engaging with 'big data' provided real-time insights on changes in mental wellbeing, behaviour and confidence in government decisions – and has also proved invaluable in helping society to build back post-COVID.[16]

---

14 Lenihan & Witherspoon, S. (2021) The place to be: how social sciences are helping to improve places in the UK, London: Academy for Social Sciences/Sage.
15 Wilsdon, J., Weber-Boer, K., Wastl, J. & Bridges, E. (2023) Reimagining the recipe for research & innovation: the secret sauce of social science, London: Sage/Academy of Social Sciences.
16 Fancourt, D.(2022) Tracking the psychological and social waves of the pandemic: the COVID-19 Social Study, Academy of Social Sciences website.

Social science research and future social challenges

On coming into office in July 2024, Keir Starmer described his government's inheritance[17] as "not just an economic black hole, [but a] societal black hole". We in the social sciences stand ready to play our part in addressing these challenges.

Social science has been critical in tackling hate crime (see case study, p29) by understanding the impact of such crimes on victims, communities and wider society, as well as identifying what can be done to lower the incidence of hate crime.[18]

On climate change, as one recent report highlights,[19] tackling net zero and building environmental sustainability will require a social mandate, which social scientists are uniquely placed to establish.

Social sciences are also vital for managing the implications of the UK's ageing population. Research from the University of Edinburgh[20] has examined the ramifications of extended working life policies, identifying complex challenges for both employers and older workers. Insights from those studies directly informed policies to improve the labour market position of older workers and raised the profile of age inclusion in the diversity agenda of employers.

[17] UK Government (2024) Keir Starmer's speech on fixing the foundations of our country: 27 August 2024, UK Government website.
[18] Academy of Social Sciences (2024) We Society podcast, S6,Ep7: Stopping hate crime, with Matthew Williams and Neil Chakraborti, AcSS website.
[19] Pennington, C. & Curtis-Kolu, T. (2024) Governance for net zero, London: British Academy.
[20] REF 2021 Impact Case Study Database (2021) Addressing the challenges associated with Scotland's ageing workforce through the development of Scottish Government policy and employer practices, REF 2021 website.

Case study: HateLab

Social science is making a difference to how hate crime is tackled. HateLab is a global data hub looking at hate speech and crime and has directly contributed to the Welsh Government's initiative Tackling Hate Crimes and Incidents – A Framework for Action and been directly embedded into the UK National Cyber Hate Crime Hub. Nisien.ai is the HateLab spin-out which uses the latest AI algorithms to accurately detect and classify online harms across platforms in real time and provides users with guidance on how to respond and counter online harms effectively.

The provision of the HateLab platform, co-created with the Hub, has fundamentally changed the way we monitor the spread of hate speech during national events.

National Online Hate Crime Hub

Image used under license from Shutterstock.com

Case study: National Living Wage

Social science is <u>making a difference to wage inequality</u>. Social science research, in collaboration with the Low Pay Commission, on wage inequality and the gender pay gap directly informed the introduction of the National Living Wage in 2016, increasing the wages of over 1.5 million workers.

Already in 2019, the main adult minimum wage rate is estimated to have been 7% higher than if the NLW policy had not been introduced. This has a particularly big impact on women, who comprise three in five workers paid at (or below) the minimum wage.

Chief Executive Officer and Director, Resolution Foundation, 22 July 2020

Image used under license from Shutterstock.com

Principle 5: The social sciences facilitate innovation

Overview

Social science is essential for transforming innovative ideas into commercially successful products and services that benefit society. These products or services can emerge directly from social science research itself or from the insights that social science provides, such as an understanding of markets, consumers, behaviour and law, which enables innovation in other sectors to be adopted, develop and flourish. In either case, social scientists' understanding of society, economics, market research, places, psychology and risk management all plays a significant role in enabling commercially successful innovations to thrive.

Innovation is often understood in terms of inventions and technology, primarily taking place within STEM. However, innovation is broader and can also come in the form of innovation in public policymaking, organisational and institutional innovation, social entrepreneurship, and (as illustrated in the second case study overleaf) data provision. In this sense, social science research can enrich and inform society to create the context in which policy, technological and community-led innovations can flourish.[21,22]

The need for a strong social science perspective on pressing problems that have historically been conceived in technological and natural scientific terms has been recognised in the move to interdisciplinarity … [It is] impossible to understand – much less to steer – the dynamics of technological development, business innovation or governance without attention to social relations, practices and processes.

Directorate-General for Research & Innovation, European Commission

[21] Wilsdon, J., Weber-Boer, K., Wastl, J. & Bridges, E. (2023) Reimagining the recipe for research & innovation: the secret sauce of social science, London: Sage/Academy of Social Sciences.
[22] University of Oxford Social Sciences Division (2021) What is social science innovation?, University of Oxford website.

Image on p31 used under license from Shutterstock.com

Enabling targeted, effective and innovative policy interventions

The social sciences are foundational to successful policy innovation. For example:

- A 2023 report[23] from the Organisation for Economic Co-operation and Development (OECD) highlighted how the adoption of a behavioural-insights approach by the UK's Foreign, Commonwealth & Development Office combined findings from STEM and social science disciplines to enable civil servants to be more adaptive and develop behaviourally informed interventions.
- The City-REDI group and its work in the West Midlands[24] has brought about innovation by introducing a spatial lens to policymaking in Whitehall by successfully advocating for 'place' to be added into the UK Government Green Book that provides the framework for policy formulation and evaluation.
- The Cabinet Office is applying a 'test-and-learn' mindset to UK Government programmes on issues such as family support, housing and skills. Much of this will be social science led, drawing together insights from frontline expertise with policy officials and digital experts to diagnose problems quickly, and experiment with and scale up solutions.[25]
- Social science insights are vital for the transition to net zero, helping governments build and maintain the social mandate necessary for large-scale societal changes and ensure interventions are place-sensitive.[26]

Driving private-sector innovation

Social science also plays an important role in private-sector innovation. In AI integration, social science research and methodologies are helping to solve complex regulatory challenges. In addition to the case study (p35), University of Essex academics from across criminology, political science and law are working together[27] to help businesses and governments balance human rights implications of AI with commercial considerations.

---

[23] Kumpf, B. & Jhunjhunwala, P. (2023) The adoption of innovation in international development organisations, OECD Development Co-operation Working Papers, No. 112, Paris: OECD.
[24] REF 2021 Impact Case Study Database (2021) Building economic resilience: UK regional and national responses to Brexit and COVID-19, REF 2021 website.
[25] Diamond, P. (2025) From instruction to innovation: delivering Labour's Five Missions in 2025, Progressive Britain website.
[26] Climate Change Committee (2024) 2024 progress report, London: Climate Change Committee.
[27] REF 2021 Impact Case Study Database (2021) Operationalising human rights standards in the governance of state and business use of data analytics and artificial intelligence, REF 2021 website.

Elsewhere, the UK is delivering strong research on digital health technologies. The LifeGuide platform, developed at the University of Southampton,[28] has allowed hundreds of healthcare practitioners with no experience of programming to develop and trial digital behavioural change interventions. Equally, research at the University of Bath[29] has improved the prevention and treatment of adolescent self-harm through the creation of the first self-harm prevention app.

Lastly, social science plays an integral role in addressing the challenges emerging from our shifting information culture. Research[30] from Cardiff University's Crime & Security Research Institute has produced new evidence and insights about how media and social media coverage can increase the public harms of terrorism, and what works in mitigating such effects. Social science evidence[31] has also featured prominently in discussions about disinformation, misinformation and harmful algorithms within social media and other platforms.

Recommendations

We welcome the recent call by the Economic & Social Research Council (ESRC) for ideas which could support major future social science data infrastructures. Further funding streams with these aims in mind would be beneficial, particularly if they provide targeted support for the ESRC to act as a facilitator and amplifier of social science in cross-disciplinary contributions to the wider UK research and innovation system.

[28] REF 2021 Impact Case Study Database (2021) LifeGuide: developing internet-based support for healthcare, REF 2021 website.
[29] REF 2021 Impact Case Study Database (2021) The development and evaluation of a self-help app (BlueIce) to prevent adolescent self-harm, REF 2021 website.
[30] Cardiff University (2018) From minutes to months, Cardiff University website.
[31] House of Commons Science, Innovation & Technology Committee (2025) Social media, misinformation and harmful algorithms, Westminster: UK Parliament.

Case study: Making fairer decisions

Social science is <u>making a difference to the use of AI</u>. The expertise of social scientists is fundamental to ensuring new technologies can fit with the world around us and have the greatest positive impact on our daily lives. Social science research identified an incompatibility between algorithms being used and European standards. To counteract this, a tool was developed to detect discrimination in AI and machine-learning systems which aligns with the European Court of Justice's gold standard for discrimination assessment.

Using our bias test, practitioners around the world can now test for biases and make better and fairer decisions.

Professor Sandra Wachter

Image used under license from Shutterstock.com

Case study: WorldPop

Social science research is <u>making a difference by using real-world data to drive sustainable development and save lives</u>. Social science research by WorldPop produces data on population distributions and characteristics at high spatial resolutions. This data enables researchers, governments, international organisations and NGOs to target interventions and drive sustainable development across a range of areas, including health, education and disaster response.

WorldPop complements traditional population data sources with dynamic, high-resolution data from satellites, surveys and cellphones to map human population distributions at high resolution, with the ultimate goal of ensuring that everyone, everywhere is counted in decision-making.

Professor Andy Tatem

Image used under license from Shutterstock.com

Principle 6: The social sciences improve people's lives by guiding good decision-making

Overview

Social science contributes vital evidence to inform government policy which touches all our daily lives, from our education at school and the healthcare we receive to our wages at work and welfare benefits.

At a time when so much policy emphasis is being placed on addressing inequalities, boosting economic growth and identifying ways to ensure socially just transitions around climate change or emerging technologies, social science perspectives are becoming ever more important. Social scientists are also adept at working across disciplines and with STEM researchers to provide multiple, nuanced perspectives.

Shaping policy in the UK

Good social science research lies at the heart of good public policy. With their understanding of our society, economy, civic culture, places and behaviours, the UK's social scientists – in universities and the wider worlds of practice, including in national and local government – are uniquely well placed to help tackle the UK Government's five policy missions. By summarising existing knowledge and what works, and articulating the best evidence from the latest research, they can help meet the needs of a government seeking to develop policy which can drive change for the better, based on evidence-informed approaches.

> **Did you know ...?**
> Around 80% of the UK Government's stated Areas of Research Interest (ARIs) relate either wholly or significantly to the social sciences.[32]

Furthermore, cross-cutting missions are likely to place new demands on joining up and engaging with multidisciplinary evidence across a wide range of disciplines and sectors. Social scientists are also adept at working across disciplinary boundaries to tackle interconnected societal problems[33] – an inherent requirement of the new UK Government's 'mission-led' approach to policymaking.

Our recent report, Beyond the Ballot,[34] illustrated many examples of social

---

[32] Academy of Social Sciences (2023) New government database encourages researchers to inform policy, Academy of Social Sciences website.

[33] Wilsdon, J., Weber-Boer, K., Wastl, J. & Bridges, E. (2023) Reimagining the recipe for research & innovation: the secret sauce of social science, London: Sage/Academy of Social Sciences.

[34] Bridges, E., Grundy, S. & Flach, T. (2024) Beyond the ballot: social science insights on eight key policy challenges, London: Sage/Campaign for Social Science.

Image on p37 used under license from Shutterstock.com

science research in the UK helping to define and diagnose the major challenges facing society, identify the policy levers to address them, and help inform interventions to implement change. The report sought to foreground social science perspectives, research and evidence relevant to public policy in the run up to the UK General Election across eight major policy themes: Health & Social Care; Inequalities & Welfare; Housing; Macroeconomics; Regional Equity & Growth; Borders & Migration; Knowledge & Technology; and Energy & Climate. The eight themes drew on a wide range of social science methodologies (quantitative, qualitative, longitudinal, field surveys, case research, action research and others) which grapple with unravelling causes and effects in multivariate and dynamic systems. The report gathered evidence-informed insights from over 100 of the UK's leading social scientists.

Principle 2 (on the UK being a social science powerhouse) set out research carried out last year[35] analysing impact case studies from the 2021 REF exercise – which in essence showed that some of the most complex and pressing problems facing society are being tackled using research which draws heavily on social science insights. Case study examples from the same report provide further detail.

Shaping policy globally

The social sciences also play a vital role in shaping policy globally. The 'Cash plus care' case study (p41) gives just one instance – others from many examples available include:

- A team of UK social scientists[36] developed a new technique for mapping corporate structures, which has been foundational to the OECD's work on tackling illicit finance in the energy trading sector. It has also informed the Extractive Industry Transparency Initiative's (EITI) guidelines on global governance of oil, gas and mineral resources. Separately, social science researchers have been instrumental in supporting OECD countries and central banking organisations anticipate and prepare for the implications of ageing populations for innovation and economic slowdown.

---

[35] Wagner, S., Rahal, C., Spiers, A. et al (2024) The SHAPE of research impact, London: British Academy.
[36] REF 2021 Impact Case Study Database (2021) Demographic structure and economic trends: planning for Europe's financial future, REF 2021 website.

- Researchers at the University of Cambridge[37] developed a novel psychological intervention to inoculate the public against 'fake news' via an interactive game in which exposing players to a weakened dose of misinformation enables them to gradually develop cognitive resistance, improving participants' ability to detect real-life instances of fake news. This has led to the subsequent development and release of anti-misinformation interventions in the UK, France, Germany, Jordan and the US.

Recommendations

To fully capitalise on the social sciences' potential to support effective policy design and implementation, we recommend that the UK Government Office for Science and the Department for Science, Innovation & Technology (DSIT) develop integrated strategies for engaging with all sectors and disciplines. A Social Science Framework, similar to the 2023 Science & Technology Framework, would allow for social science insights to contribute more directly to the UK Government's technological, social, economic and environmental priorities.

A social sciences ARIA could be tasked and funded to identify new ways of tacking the big social and economic issues facing the UK today.

---

[37] REF 2021 Impact Case Study Database (2021) A psychological vaccine against fake news, REF 2021 website.

Case study: Cash plus care

Social science research is <u>making a difference to AIDS-affected children</u>. Research on children and young people affected by AIDS in Africa has led to major systemic social change at the interface of health, poverty, education, behaviour and policy, with 'cash plus care' services being found to substantially reduce the risk of HIV infection in boys and girls. These services have been delivered to over 2.5 million children in over ten African countries.

The concept [of 'cash plus care'] has become an integral part of UN planning and response to the HIV epidemic.

Senior Advisor,
UNAIDS

Image used under license from Shutterstock.com

Case study: CAST

Social science is <u>making a difference by putting people at the heart of climate change action</u>. Social science researchers have been instrumental in exploring and explaining how human behaviour might need to adapt to slow climate change. The Centre for Climate Change & Social Transformations (CAST) has highlighted the importance of a people-centred approach to climate action. Its work is helping to embed climate action within local government decision-making and contributing to the Skidmore Review of Net Zero.

We want to work closely with people and organisations to achieve positive low-carbon futures – transforming the way we live our lives, and reconfiguring organisations and cities.

Professor Lorraine Whitmarsh

Image used under license from Shutterstock.com

Principle 7: Social science subjects are open and available to all

Overview

The UK Government has made clear its commitment[38] to building skills and opportunities so that every young person can follow the right pathway for them. Principle 8 in this report sets out how and why the social sciences are a popular route for many young people. This section examines the diversity and inclusivity of social sciences.

Social science disciplines offer excellent equality of opportunity nationally, regionally and locally, a critical contributing factor to our disciplines' accessibility. Not only do our universities make the UK a social science learning powerhouse (see Principle 2), there are also many opportunities within Further Education for vocational training in the social sciences.[39] These can include practical and supporting roles in areas such as law (e.g. legal assistants), social welfare (e.g. care workers) and finance (e.g. accounting assistants).

At the Higher Education level, analysis of data from the UK's Higher Education Statistics Agency[40] indicates that the social sciences perform well across the four main EDI characteristics (sex, ethnicity, nationality and disability). For some groups and characteristics, the social sciences are broadly in line with other academic sectors. For other groups, the social sciences show higher levels of diversity when compared to 'all HE' HESA data. Nevertheless, there are areas for further improvement and exploration.

Did you know ...?

- 58% of social science students at UK universities are female.
- 25% of UK social science students of a known ethnicity at university are from ethnic minority groups, including 8% Black and 12% Asian.
- 14% of social science students at UK universities have declared a disability.

---

[38] UK Government (2024) Break down barriers to opportunity, UK Government website.
[39] Academy of Social Sciences (2024) Academy of Social Sciences' response to the Curriculum and Assessment Review for England, AcSS website.
[40] Higher Education Statistics Agency (2023) Higher Education Student Statistics: UK 2021/22 release, HESA website.

Image on p43 used under license from Shutterstock.com

The analysis is summarised in <u>our recent report</u>, which forms part of our wider collaborative EDI Project in partnership with our member Learned Societies and the ESRC.

Social sciences academic staff

There are higher proportions of women in senior management positions (45%) and professor roles (36%) within the social sciences than across academia as a whole (42% and 30% respectively), although there remain twice as many male professors within the social sciences sector.

Figure 1 shows there are higher proportions of all ethnicities in permanent/open-ended contracts in the social sciences than across academia as a whole, with those of Asian and White ethnicity having the highest proportions across the social sciences (with 82% of Asian staff and 82% of White staff being in permanent/open-ended contracts).

Figure 1: Proportion of academic staff in permanent / open-ended contracts by ethnicity

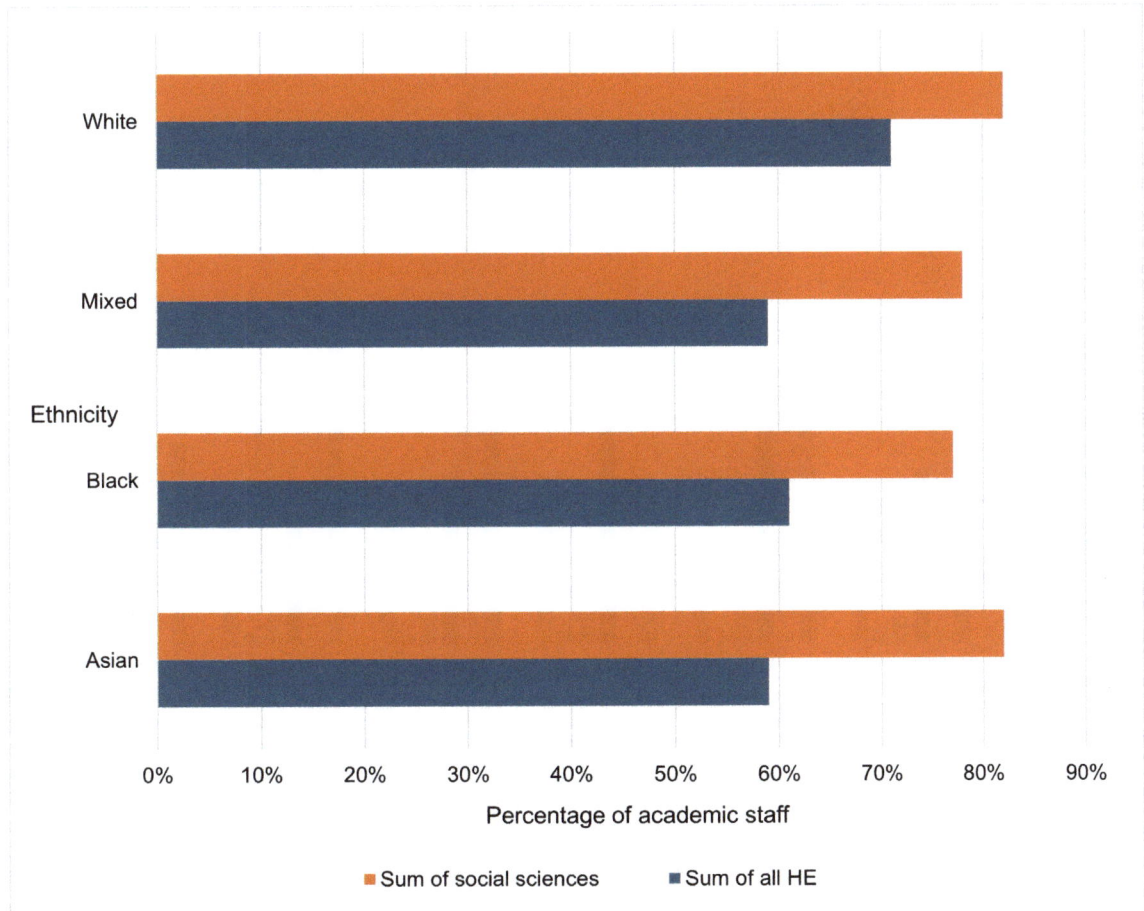

Social science students

More women (58.2%) than men (41.8%) study social sciences, which is in line with sector-wide data (57% women, 43% men across all disciplines). At undergraduate level, a higher percentage of women are awarded first-class honours degrees than men (26.7% to 22.5%).

The ethnic breakdown of the student population within the social sciences (71% White, 12% Asian, 8% Black, 5% Mixed) is broadly in line with the wider academic sector. However, Black students are less likely to receive first-class and upper second-class degrees than other ethnic groups, whilst a higher proportion of White students are awarded first-class degrees compared to the total social science undergraduate population. As Figure 2 shows, 12.4% of Black students receive a first-class degree, compared with 25% across all social science undergraduate students, and 28.1% of White students.

Figure 2: Ethnic breakdown by degree-awarding class in the social sciences (undergraduate students only)

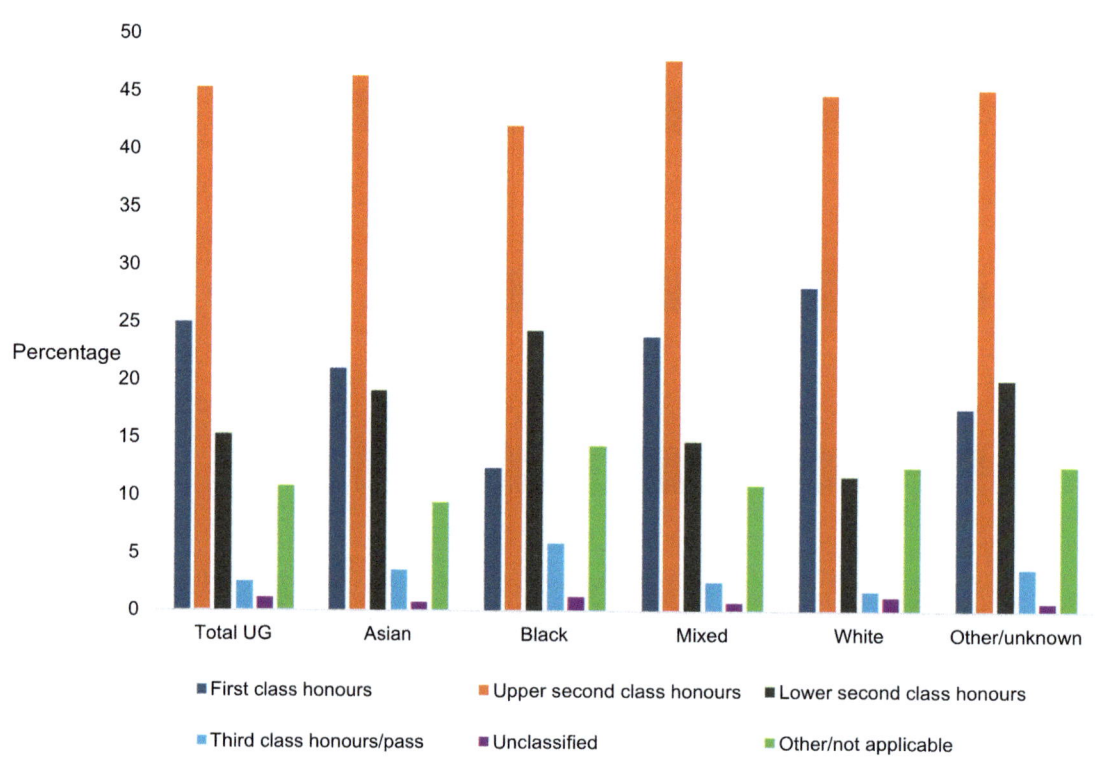

The social sciences also differ from the total student population, with a higher proportion of social science students coming from the upper quintiles (more advantaged) than the total student population data. This will have complex interactions with the other variables, affecting the data.

How we will improve things further
To boost EDI across the social sciences, we are working with the ESRC and some of our member Learned Societies on a range of action-focussed projects. Details can be found on our website.

The skills you need for this role include strong analytical abilities for evaluating financial data, excellent communication skills for explaining complex ideas, problem solving and critical thinking skills for navigating challenges, and interpersonal skills for dealing with the human aspects of mergers and acquisitions.

James Oyuke
Director of Mergers and Acquisitions, private sector

While you might specialise in a particular social science, reading into other social sciences and engaging with other social scientists can really help you – people and society are complicated, so we need to work together as the social science community to understand them as much as possible.

Kiran Krishna
Principal Economist,
Information Commissioner's Office

Principle 8: Social science skills are valued by employers and lead to well-paid careers

Overview

Social scientists have skills and knowledge that are in demand across a range of sectors leading to a variety of interesting, rewarding and well-paid careers. The income of social science graduates is in line with average expected graduate earnings and, in some cases, significantly more than the average graduate salary, with economics, law, politics, business studies and geography being among the top ten disciplines for lifetime graduate earnings.

Three key messages

Firstly, **the social sciences are popular and equip students to understand the contemporary, fast-changing and complex human world in which they live and the challenges and opportunities that face it**. Social science skills are an essential aspect of a broad and balanced education, preparing young people

> ### Did you know ...?
> Over half of MPs elected in 2024 have either a degree or Master's in a social science subject, and almost three quarters of the current UK Government cabinet have a social science background.

for their lives ahead by providing insights into human behaviour, societies, economies and places. They also encourage critical thinking and analysis by examining social phenomena and contemporary issues from multiple perspectives, as well as providing a framework for broader civic engagement by equipping students with knowledge about governance, democracy and economic systems, empowering them to become informed, engaged citizens.

- 47% of UK students graduate from university with a social science degree,[41] reflecting not only the popularity of the 15 or so social science subjects because of their content, but also the fact that students consider these subjects the key to engaging and rewarding careers.[42]
- Not only are social science disciplines extremely popular, but they are particularly important to women, especially in subjects such as law, education, sociology and (social) psychology, where the proportions of female students are significantly higher than their male counterparts.[43]

[41] Higher Education Statistics Agency (2023) What do HE students study?, Cheltenham: HESA.
[42] Academy of Social Sciences (2023) Careers for social scientists briefing note, AcSS website.
[43] Higher Education Statistics Agency (2023) What do HE students study?, Cheltenham: HESA.

In addition, **the social sciences deliver skills which are in high demand by employers, allowing students to access a wide range of employment sectors and contribute to the economy and society**. Social scientists work across all industries and sectors, deploying a variety of skills to the benefit of society. Some social science professions, including law, accountancy and planning, are closely linked to particular subjects and require specialist qualifications. Other roles relate directly to the knowledge and skills specific to a subject, for example, geospatial analysts trained in geography, operations managers trained in business and management, economists working in the NHS as programme evaluators, political scientists working for polling companies, and psychologists working on organisation behaviours. Many other jobs and careers are open to all social scientists, making use of the wide range of transferable skills gained by studying a social science degree.[44]

- According to the latest Longitudinal Education Outcomes (LEO) data,[45] on average, 87% of UK-domiciled first-degree graduates across the social science disciplines were in sustained employment, further study or both five years after graduation. This was slightly lower than the average (89%) for STEM graduates, and higher than the average (84%) for arts and humanities graduates.
- Research from the British Academy[46] and others has shown that UK social science graduates possess a set of transferable skills, or core skills, that employers find valuable. These include the ability to communicate clearly and work effectively with others, the capabilities to design research, collect and analyse evidence and data, and to hone useful skills in problem-solving, independence, flexibility, creativity and adaptability. Meanwhile we have set out previously[47] that numeracy and data skills are an intrinsic part of social science degrees but are often not recognised as such.

[44] Greaves, L. (2024) What do graduates do? Insights and analysis from the UK's largest higher education survey, Bristol: Prospects Luminate.
[45] Department for Education in England (2023) LEO graduate outcomes provider level data (tax year 2020/21), UK Government website.
[46] British Academy (2017) The right skills: celebrating skills in the arts, humanities and social sciences, London: British Academy.
[47] Academy of Social Sciences (2024) Academy of Social Sciences' response to the UK Government Curriculum and Assessment Review for England, AcSS website.

Finally, **social science students go on to have good earnings potential, comparable to those in STEM fields**. Some of the social science professions – not least law, accountancy and marketing – are at the heart of all business operations in UK public limited companies. Furthermore, the UK's business schools in universities – also a constituent part of the social sciences – train thousands of people each year from across the world in management skills; and economics, statistics, geography and social research skills support public sector analysis across UK and devolved governments. As a result:

- Five years after graduating, the median salary for a social science first-degree graduate was £28,438 (tax year 2020/21). This compares to an equivalent arts and humanities figure of £23,329 and an equivalent STEM figure of £31,129[48] (which rises to £33,200 with medicine and dentistry included).
- Five social science disciplines are among the top ten disciplines for lifetime graduate earnings: economics, law, politics, business studies and geography. Social science graduates as a group fare about as well as graduates from STEM disciplines in terms of lifetime earnings.[49]

[48] Department for Education in England (2023), LEO graduate and postgraduate outcomes (tax year 2020/21), UK Government website.
[49] Britton, J., Dearden, L., der Erve, L. & Waltmann, B. (2020) The impact of undergraduate degrees on lifetime earnings, London: Institute for Fiscal Studies: pp40-41.

> **If you're curious about the world, the actors and institutions in it, social science allows you to start probing, asking and interrogating those things. Having an understanding of what makes humans tick, how we make decisions, understanding research design and methods, good and poor-quality evidence and how to use research and data in practice, will stand you in good stead for many career options.**
>
> Ellie Brown
> Director of Impact,
> Get Further

In an increasingly complex world, the social sciences are one of the principal ways to cut through the noise and begin to navigate and offer ways to frame problems and pathways to solving them. If you want to stand out in a job market, especially in a context with the increasing role of AI, then studying the social sciences can help set you apart.

Alex Clegg
Local Economic Growth Policy Advisor,
Civil service

Drawing it all together: conclusions and recommendations

The world is in the grip of three 'revolutions' affecting our societies, economies and places, and presenting new challenges for us to grapple with and opportunities for us to grasp, both now and over the coming decade.

The first is the collective human impact on our planet encompassing climate change, pollution, biodiversity loss and other environmental impacts. This not only underpins calls for sustainability but has also led to growing global pressures of displacement and migration, poverty and hunger, and disease and conflict.

The second is the rapid and globalised shift in technology use, with AI and machine learning unlocking great economic opportunities, but also posing major societal threats and demands for governing frameworks.

Thirdly, we are on the cusp of great advances in biomedical research that pave the way for better managing chronic illnesses and extending life spans, whilst bringing increasingly difficult decisions of cost, access and equality in public health service provision in an already resource-constrained world. All of which is taking place within increasingly uncertain geopolitical contexts.

The social sciences sit at the heart of understanding the differentiated and dynamic impacts (both positive and negative) of the revolutions on people, economies and environments, and in helping to inform and influence governmental and societal responses to them.

In the UK, we have not only these forces at play, but also the legacies of past political ideologies and decisions and a context of political populism. We are at a critical time in the life of the nation.

All the challenges, and the opportunities they offer, are inextricably linked to both science and technology and, symbiotically and often in multiple ways, to people, societal structures, economies, behaviours and governance, locally and globally. In other words, they are firmly linked to the social sciences. Impacts on people are often differentiated by complex interactions of location, socio-economic characteristics, ethnicity, levels of education, community structures and behaviours, and require rigorous analysis in the social sciences to understand them. In short, the social sciences are not simply helpful but are essential to improving the lives of people in the UK given the breadth, depth and interconnectedness of the challenges and opportunities we face.

This report demonstrates the eight broad ways in which the social sciences are contributing to the wellbeing of the UK and its devolved nations and regions during this vital period. They are performing well: delivering world-leading enquiry-driven and applied research using robust approaches; equipping the next generation with vital skills and good employment prospects; and supporting policymaking and communities across the UK's nations and regions – but they could be doing more.

To facilitate this, the Campaign for Social Science argues that two supporting conditions need to be fulfilled:

1. **Adequate and long-term funding of social science research needs to be secured**. This is not simply a 'more money' call, and we are conscious of the many demands on government spending at the current time. Nevertheless, as this report sets out, the social sciences are increasingly expected to deliver world-class research on budgets which are anything but. Better-targeted funding of the social sciences, and a more sympathetic funding landscape, should be seen not as a cost to government but an investment in a sector which is already delivering for UK society and people, and which will continue to repay that investment through the contribution and impact it makes to the world around us.

   We have identified in the preceding pages some of the mechanisms which would help to redress the balance. These include:
   - increasing QR funding which is essential to our disciplines' sustainability;
   - including social science-based R&D within the tax relief framework;
   - expanding the Frascati rules for measuring R&D so that they reflect the social sciences' contribution;
   - reflecting more fully the social sciences' involvement in multidisciplinary, challenge-led research; and
   - considering specific additional funding for education research, for research that benefits specific UK nations/regions, and research in the social aspects of health sciences because of their centrality to delivering on the UK Government's stated policy objectives.

2. **A structural step-change is made within the UK Government and its devolved counterparts to harness the social sciences' unmet potential.**
Governments at all levels in the UK would stand to receive much more benefit if they better recognised and reflected the strength in social sciences, beyond economics, at their disposal and embedded it fully within their science, technology and evidence systems. And too often the many social scientist graduates in government are shy about speaking up for their disciplines and the skills with which those disciplines have equipped them. This simple act could do so much for raising the understanding and attention paid to making the most of what the social sciences have to offer.

We have set out in this report some examples of how that might be addressed. For example, the UK Government's Office for Science could develop integrated strategies for engaging with all sectors and disciplines, whilst the Department for Science, Innovation & Technology could develop a Social Science Framework, similar to its Science & Technology Framework, so that social science insights can contribute more directly to the nation's technological, social, economic and environmental priorities.

The Academy of Social Sciences will play our part in better harnessing the potential of the social sciences via our project to understand how scientific evidence – including that from the social sciences – can be used better in policymaking and implementation, across UK Government, to help provide better outcomes for citizens. We anticipate releasing formal recommendations before the end of 2025.

We will also continue to act as a gateway to leading expertise in the social sciences through our Fellows and our ability to convene expert groups and co-ordinate regional groups of expertise, as shown most recently in our report In the region, for the region: the Midlands. The latter includes some of the best examples of collaborative applied social science research supporting the recent economic development of the Midlands region.

The eight principles exhibited in this report will also continue to inform the Academy's broader work to increase public understanding of the social sciences and their contribution, to the extent that our resources allow. This includes through channels such as our We Society podcast series to showcase how social science is shaping the ways we live, through our careers hub, which identifies the skills and opportunities provided by education and training in social sciences, and our I am a Social Scientist series, featuring social scientists from a range of backgrounds discussing their career paths and what inspires them in their work.

Finally, we call upon others to play their part too. While we are the only body that solely represents the social sciences sector, others also play vital roles. This includes the UKRI and independent research funders in terms of allocation of funds, the universities in supporting and building their strengths in the social sciences in the current difficult climate they face, the relevant learned societies and the British Academy in facilitating and demonstrating the role of the social sciences (and humanities) in helping to build a better future in a time of great change. And social science graduates, academic communities and professional communities are encouraged to be confident in identifying themselves as social scientists, as well as topic or subject experts, as a simple, pragmatic and pervasive way of demonstrating the contemporary relevance of the social sciences.

Dr Rita Gardner CBE FAcSS
Chief Executive
Academy of Social Sciences

Image courtesy of Sandi Friend.

Acknowledgements

The Academy of Social Sciences would like to thank Dr Ed Bridges, Head of Policy and Public Affairs, and Amy Williams, Senior Communications Manager, for their significant contributions to this report, as well as to other members of the Academy's team who supported its production.